THE BRITISH PLANTATIONS
ASSOCIATION OATH ROLLS.

— *1696* —

THE VIRGINIA ROLLS, PAGES 30-32.
Reproduced by kind permission of the Master of the Rolls.

THE ASSOCIATION OATH ROLLS
OF THE BRITISH PLANTATIONS
[New York, Virginia, Etc.] A.D. 1696.

Being a Contribution to Political History

EDITED with an INTRODUCTION

BY

WALLACE GANDY

Editor of "The Lancashire Oath Rolls," Author
of "Guide to some Original Manuscript Sources
of British and Colonial Family and Political
History," "In the Days of Lionheart," etc.

CLEARFIELD

Originally printed 1922

International Standard Book Number 0-8063-4529-2

Reprinted for
Clearfield Company, Inc. by
Genealogical Publishing Co., Inc.
Baltimore, Maryland
1993

PREFACE.

NO apology is neccessary for the issue of those heretofore inedited sources of English and Colonial history of that absorbingly interesting and pregnant period, the reign of William III. The Rolls have been transcribed *verb. et lit.* and with the Introduction, will form an indispensable guide to a fuller understanding of the political problems of that day, so fruitful of historical suggestion shedding further light on the problems of to-day.

As Professor Murray Butler has well said, " there is a deep and ineradicable connection between the free institutions of England and America," and I would add, the Britains Overseas.

It is my hope that this book will find its way into every educational institution where work is done by the ever-increasing number of students of Anglo-American relations, to the end that the essential unity of spirit of the peoples of our race may be better understood.

In conclusion I must avow my indebtedness to Professor A. P. Newton, who has kindly read the proofs; the Rev. J. Harvey Bloom, M.A., who has assisted me with the transcription; and the Officials of H.M. Record Office for generous help and advice.

W.G.

TO

NICHOLAS MURRAY BUTLER,

PRESIDENT OF COLUMBIA UNIVERSITY

AND MEMBER OF THE AMERICAN ACADEMY OF ARTS AND LETTERS

Eminent alike as man, thinker, and scholar.

A token of regard.

CONTENTS.

The Association Oath Rolls for the British Plantations. A.D. 1696.

HE reigns of William and Mary, and of William III. alone represent a period of the greatest importance to the student of the development of political machinery. It was essentially the epoch marked by the introduction of a stringent parliamentary control of finance, of party government, the appointment of Ministers, and consequent introduction of the Cabinet, leading in the next reign to the establishment of a cabinet under a Prime Minister definitely charged with the task of governing. With the exception of George III. William was the last British sovereign who sought to rule as well as to occupy the throne.

His reign was greatly troubled, as was natural, by plots in favour of the exiled monarch, James II., the effect, however, being to enhance the security of William's seat. In spite even of the tenacity of Parliament in wresting from the crown every power they could, the popular esteem of the new king continued to increase. It was strikingly manifested early in 1696, immediately after the king made his dramatic announcement in Parliament that a plot against his life had been discovered.

The Whigs had materially assisted his accession to the throne and the Tories, smarting under the attacks upon the Church of England made by James, had taken their part in the successful penetration of England by the Prince of Orange. There was indeed considerable connection between the religious and political movements of the time. The Catholic party in European politics were greatly pleased at two notable events in the year 1685. James, Duke of York, ascended the throne of England an avowed papist, and in France the Edict of Nantes was repealed on October 22nd, in part for political, in part for religious, reasons by Louis XIV. The repeal was followed by such cruelty and bitter persecution that the Protestant conscience was awakened as it had never been before. Their imminent danger evoked general alarm, as they witnessed the hunting down and dispersion of their Huguenot brethren. It made such an appeal to their religious sentiment and the call for self-preservation that its effects are still found, even in these days when all sectarian bitterness is to some extent forgotten.

The year 1689, which saw James II. a fugitive, witnessed the outbreak of the Palatinate war and the beautiful valley of the Rhine was ravaged. In her fear England called William of Orange to the throne to be both the defender of the Protestant cause and joint ruler, with Mary his Queen, of Great Britain. As a man he was neither loveable nor loved, his melancholy nature permitted him few friends. Yet, none the less, he was an able monarch and a commander of ability.

The Revolution was sufficiently remarkable. The Convention Parliament, composed of all surviving members of the Commons under Charles II., augmented by the Lord Mayor, aldermen and fifty common councillors of London, the bishops and peers, declared the throne to be vacant, James having " broken the original contract between King and people," having violated the fundamental laws and having abdicated the government by withdrawing himself out of the Kingdom.

In other words, the Sovereign people passed judgment, and in their new-found power elected their king, taking care by a cautious settlement to throw aside the burdensome yoke of the ill-fated Stuarts, and form such a contract that arbitrary government could not again endanger the liberty of the subject. Among the earliest Acts of Parliament was one calculated to ensure the Protestant succession, and naturally peace with Louis, as protector of the exiled Catholic king, was out of the question. The liberty, laws and religion of England were at stake. James was not however without support.

Many of the landowners at that time as did their forebears, owned Papal allegiance, and looked with lenient eye on the humbler estrays of the faith in their midst, whether Irish or English, welcoming them none the less if they had chanced to serve under the banner of the late king.

His co-religionists were with James to a man, and outside their pale many still looked with awe on the doctrine of Divine Right of Kings, and were hostile to the Dutchman. Ireland was ready and willing to fight, and it was only after much bloodshed brought to order, with William and Mary securely on the throne. Plots were fomented and means sought to overthrow them. Lancashire was conspicuous in these machinations, but it by no means stood alone.

Charles Talbot, Duke of Shrewsbury, one of those who had escorted the Prince of Orange from Holland in 1688, pointed this out to the justices as early as April, 1690, in a letter written in his official capacity as Secretary of State for the Northern Province, and referred again to the matter in another letter to William, dated July 10th, 1694.* In the latter, he says :—

> Three Lancashire and Cheshire men, who have been engaged enlisting soldiers, and buying arms for the disaffected gentlemen in those parts, have made an ample discovery of the whole matter to the Lord Keeper, Mr. Secretary and myself. Whereupon Mr. Secretary and I have sent out warrants for seizing the persons and arms of about twenty considerable gentlemen thereabouts, and if these witnesses make good at their trial what they have deposed before us, they will be every man both lives and fortunes, in your power. [S.P. Dom. 1694—5.]

This so-called " Lancashire Plot " was probably made to look as black as possible. It cannot have been very serious. Nevertheless, the principal organisers, Lord Molyneux of Croxteth, Sir William Gerard, Sir Rowland Stanley, Sir Thomas Clifton, Bartholomew Walmsley, William Dicconson, Philip Langton, Esqs., and Mr. William Blundell—all but Stanley being men of the County—were duly brought to trial for high treason.

The Duke's informers, however, hardly fulfilled his expectations, and in the upshot all those charged were acquitted with a severe censure.

According to the Court scribe, Richard Kingston, Lancashire was not only the parent but the companion, of all other conspiracies contrived by the exiled King and his adherents. This Lancashire plot was therefore one of many.

A more serious conspiracy was formed in 1695, but its execution was postponed owing to the king's departure for the Continent; but after his return it revived, and came to a head in February, 1696, when it was disclosed to William Bentinck, Earl of Portland, by Captain Fisher (February 11, 1696). Portland undoubtedly handled the conspiracy with consummate skill in

*See Introduction to Lancashire Association Oath Rolls. Pub. 1921.

William's interests. This new venture was more or less indirectly and somewhat vaguely supported by general sanction from the Court of St. Germain. Sir George Barclay, a Scottish Catholic, acted as agent for the exiled James, and was commissioned to rally his adherents. James, Duke of Berwick, a son of James II., had already secretly arrived in London, as agent for Louis, who on his part, prepared a French fleet at Calais and Dunkirk. James II. waited at Calais to take command. Sir John Fenwick led the conspiracy, in which about forty plotters were involved.

William was in the habit of hunting in Richmond Park on Saturdays, and it was decided to make an attempt on his life in the lane leading from Brentford to Turnham Green. In spite of warning, William persisted in his intention up to the evening of Friday, February 14th, when Prendergast, an Irish Catholic, called upon Portland, who saw the King, with the result that the hunt was postponed. On the following day, Prendergast and another witness, De la Rue, were questioned by the King, and they revealed the names of the conspirators.

Several of the plotters were arrested that night, and rewards of £1,000 were proclaimed for the capture of the others. One such Proclamation is reprinted, the text being quoted from the authorised report of Rookwood's trial, printed for Samuel Heyrick and Isaac Cleave, 1696. As was pointed out during that trial the lists of names varied slightly.

THE PROCLAMATION.

By the King a Proclamation.

WILLIAM R.

WHEREAS His Majesty has received information upon Oath, that the Persons herein after named, have with divers other wicked and traiterous Persons entered into a horrid and detestable Conspiracy to assassinate and murder His Majesty's sacred Person, for which cause several

Warrants for High Treason hath been issued out against them, but they have withdrawn themselves from their usual places of abode, and are fled from Justice: His Majesty has therefore thought fit by the advice of his Privy Council to issue his Royal Proclamation, and His Majesty does hereby command and require all His loving Subjects to discover, take, and apprehend JAMES, DUKE OF BERWICK, SIR GEORGE BARCLAY, MAJOR LOWICK, GEORGE PORTER, CAPT. STOW, CAPT. WALBANK, CAPT. JAMES COURTNEY, LIEUTEN SHERBORNE, BRICE, BLAIR, DINANT, CHAMBERS, BOISE, GEORGE HIGGINS, and his two brothers, Sons to Sir Thomas Higgins, DAVIS CORDELL, GOODMAN, CRAMBURNE, KEYES, PENDERGROSS, *alias* PENDERGRASS, BRYERLY, TREVOR, SIR GEORGE MAXWELL, DARANCE a Fleming, CHRISTOPHER KNIGHTLEY, LIEUT. KING, HOLMES, SIR WILLIAM PARKYNS, ROOKWOOD, wherever they may be found, and to carry them before the next Justice of Peace or chief Magistrate, who is hereby required to commit them to the next Gaol, there to remain until they be thence delivered by due course of Law. And His Majesty doth hereby require the said Justice, or other Magistrate, immediately to give notice thereof to Him or His Privy Council. And for the prevention of the going of the said Persons, or of any other, into Ireland, or other parts beyond the Seas, His Majesty does require and command all His Officers of the Customs, and other His Officers and Subjects of and in the respective Courts and Maritime Towns and Places within His Kingdom of England, Dominion of Wales, and Towns of Berwick upon Tweed, that they and every of them in their respective Stations and Places, be careful and diligent in the Examination of all Persons who shall pass or endeavour to pass beyond the Seas, and that they do not permit any Person whatsoever to go into Ireland, or other places beyond the Seas, without a Pass under His Majesty's Royal Sign Manual until further Order. And if they shall discover the said Persons above-named or either of them, then to cause them to be apprehended and secured, and to give notice as aforesaid. And His Majesty does hereby Publish and Declare to all Persons who shall Conceal the Persons above-named, or any of them, or be aiding or assisting in the Concealing of them, or furthering their Escape, That they shall be proceeded against, for such their Offence, with the utmost severity, according to Law. And for the Encouragement of all Persons to be Diligent and Careful in endeavouring to Discover and Apprehend the said Persons, We do hereby further Declare, That whosoever shall Discover and Apprehend the Persons above-named, or any of them, and shall bring them before some Justice of Peace or Chief Magistrate, as aforesaid, shall have and receive as a Reward, the Summ of One thousand Pound; which said Summ of One thousand

Pounds, the Lords Commissioners of His Majesty's Treasury are hereby Required and Directed to pay accordingly. And if any of the Persons above-named shall Discover and Apprehend any of their Accomplices, so as they may be brought to Justice, His Majesty does hereby declare, That every Person making such Discovery, shall have His Majesty's Gracious Pardon for his Offence, and shall receive the Reward of One thousand Pounds, to be paid in such manner as aforesaid.

Given at our Court at Kensington, the 23rd Day of February, 1695/6, in the Eighth Year of our Reign.

GOD SAVE THE KING.

The principal persons implicated were the Earl of Aylesbury, Lord Montgomery, Sir George Barclay, Sir John Fenwick, Sir John Friend, Sir William Perkins, Major Lowick, Capt. Charnock, Capt. Knightley, Capt. Porter, Messrs. Cooke, Cranbourne, Goodman and Rookwood. Of these, Friend, Perkins, Charnock, Lowick, Cranbourne, and Rookwood were executed : Cook was banished. Sir John Fenwick was arrested later (June 11). Proceedings against him were begun, in the course of which there was much intrigue. A bill of attainder was introduced, and after lengthy and acrimonious debate, Fenwick was condemned to death and was beheaded on Tower Hill (January 28th, 1697).

Politically the plot was of great service to William, and, like the statesman he was, he made the most of the situation. On Monday, the 24th February, William, in person, appeared in Parliament, and made the following speech, duly reported in the *London Gazette* that day :—

I am come hither this day upon an Extraordinary occasion, which might have proved Fatal, if it had not been Disappointed by the singular mercy and Goodness of God, and may now, by the Continuance of the same Providence, and our own Prudent Endeavours, be so improved as to become a sufficient Warning to us to provide for our Security against the pernicious Practices and Attempts of our Enemies. I have received several concurring

Informations of a Design to assassinate me, and that our Enemies at the same time are very forward in their Preparations for a sudden Invasion of this Kingdom, and have therefore thought it necessary to lose no time in acquainting my Parliament with these things, in which the safety of the Kingdom and the Publick Welfare are so nearly concerned, That I assure myself nothing will be omitted on Your Part which may be thought proper for Our present or future Security. [After setting out the steps taken to bring the Navy to home waters, and a large number of soldiers home, the King concludes.] Having now acquainted you with the Danger that hath Threatened us, I cannot doubt of your Readiness and Zeal to do everything which you shall Judge Proper for Our Common Safety, and I Persuade myself we must be all Sensible how Necessary it is in our Present circumstances that all possible Dispatch should be given in the Business before you.

The Lords and Commons replied with an address declaring " detestation and abhorrence of so vilanous and barbarous a design," urging his Majesty to take more than usual care of his Royal Person, and promising not only support, but to take revenge upon his enemies should the King meet with a violent death. They empowered the King to suspend the Habeas Corpus Act. In the House of Commons Sir Rowland Gwyn proposed an Oath of Association which was drawn up and immediately signed by an overwhelming majority : 400 to 113.

The authorities of the City of London signed on February 25th, and while the Houses proceeded to pass an Act to legalise it, the oath was taken in town and country, and the South coast put in a position of defence. This Act is quoted as 7 & 8 Will. III., c. 27, and entitled "An Act for the better security of His Maties Royal Person and Government."

PREAMBLE OF THE ACT.

Rot. Parl. 7 & 8 Gul. III. p 6. n 1. 7 & 8 Will. iii c 27. An Act for the Security of His Maties Royal Person and Government.

Whereas the Welfare and safety of this Kingdom and the Reformed Religion do next under God entirely depend upon the Preservation of Your Majesties

Royal Person and Government which by the Merciful Providence of God of late have been delivered from the bloody and barbarous attempts of Traytors and other Your Majesties Enemies who there is just Reason to believe have been in great measure encouraged to undertake and prosecute such their wicked designs partly by Your Majesties great and undeserved Clemency towards them and partly by the want of a sufficient Provision in the Law for the securing offices and Places of Trust to such as are well affected to Your Majesties' Government and for the repressing and punishing such as are knowne to be disaffected to the same. For Remedy whereof may it please your Majesty that it may be enacted That from and after the First day of May, One Thousand Six Hundred and Ninety-six all and every such Person and Persons who shall refuse to take the oaths mentioned and appointed to be taken in an Act of Parliament (1 Wm. & M. c. 8) intituled An Act for the abrogating of the oaths of Supremacy and Allegiance and appointing other oaths or either of them when tendred to him or them by Persons lawfully authorized to administer or tender the same or shall refuse or neglect to appeare when lawfully summoned in order to have the said oaths tendred to him or them shall until he or they have duly taken the said oaths be liable to incur forfeit pay or suffer all and every the Penalties Forfeitures Sums of money disabilities and incapacities which by the Laws and Statutes of this Realme now in Force or any of them are inflicted upon Popish Recusants duely convict of Recusancy and that for the better and more orderly levying and answering the said Penalties and Forfeitures to his majesty His Heirs and successors the Persons so tendring the said oaths shall upon every such Refusal or Default of appearance as aforesaid record and enter in Parchment the Christian and Sirnames and the Place of abode of the Person or Persons so refusing or not appearing as aforesaid to take the said oaths or either of them together with the time of such Tender and Refusal or Default of appearance and shall deliver and certifie the said Record or Entry to the Justices of Assize, Justices of Oyer and Terminer or Gaol Delivery att their next session within the same county who shall forthwith estreate and certifye the same into His Majesties' Court of Exchequer to be there entered of Record to the end that the said Court of Exchequer may thereupon award and issue such processe against the lands and goods of the said Person or Persons mentioned in such Estreat or Certificate as may by the Laws and Statutes of the realme be awarded and issued against the lands and goods of a Popish Recusant convict. [§ I.]

The preamble is immediately followed by a section forbidding that any by writing, printing, preaching, teaching, or advised speaking, shall publish or declare that His present Majesty is other than lawful and rightfull King of these Realms, or that the late King James, or the pretended Prince of Wales or any other, have right and title to the crown, otherwise they incur the danger and Penalty of Premunire (16 Ric. II., c. 5). [§ II.]

The oath which follows below is introduced as for " the better Preservation of His Majesties Royal Person and Government," and to it was added the text of the association which had already been subscribed by " great numbers of His Majesties good subjects."

THE OATH.

Whereas there has been a horrid and detestable Conspiracy formed and carried on by Papists and other wicked and traiterous Persons for assassinating His Majesties Royal Person in order to encourage an Invasion from France to subvert our Religion Laws and Liberty. Wee whose names are hereunder subscribed doe heartily sincerely and solemnly professe testifie and declare That His present Majesty King William is rightfull and lawful King of these Realmes and wee doe mutually promise and engage to stand by and assist each other to the utmost of our Power in the support and defence of His Majesties most sacred Person and government against the late King James and all his adherents and in case His Majesty come to any violent or untimely death (which God forbid) wee do hereby further freely and unanimously oblige ourselves to unite associate and stand by each other in revenging the same upon His Enemies and their adherents and in supporting and defending the succession of the Crown according to an act made in the first yeare of the Reigne of King William and Queen Mary entituled an Act declaring the Rights and Liberties of the subject and settling the succession of the Crowne.   [§ III.]

It may be noted that Gwyn's suggestion was not a novel one.* Among the State Papers Domestic is preserved a 4-page quarto, viz., " A true copy of the instrument of association, that the Protestants of England entered into in the 27th year of Queen Elizabeth, against a Popish conspiracy." In this occur several passages, the spirit of which seems to be analogous to, and even identical with that of the later oath.

*Vide Introduction Suffolk Association Oath Roll. MS. by Mr. H. W. B. Wayman, 1909.

Throughout England the terms of the Act were far exceeded. In most places every male adult, and occasionally certain women, signed the roll either in person or by deputy. Particularly thorough supervision was exercised in obtaining the signatures of all connected with government offices and works. Nor did the movement end with the old country. The English merchants of Dort, Rotterdam, The Hague, and Geneva, contributed their rolls; so, too, the Factory at Malaga. And across the ocean the West Indies, New York and Virginia also took the oath.

Conditions in these distant plantations require some short notice in order the better to understand what the Association meant to the settlers.

The plantations on the Mainland of North America owed some gratitude to the Stuarts. James I. had been interested in Virginia, as had Charles I. Maryland began officially to exist in 1634; Rhode Island two years later; Pennsylvannia as late as 1664. James II. touched a sore spot, though his purpose was a wise one. To set aside their charters and unite the Northern Colonies under one government was far from approved of by their liberty-loving people; but since the natural jealousies of New York and New Jersey were at the time in full play, the need of some centralizing movement is apparent. All the while the French of Lousiana and Canada were aggressive. Dougan, Governor of New York, sagaciously opposed their machinations and tactfully kept the great Indian confederacy on the borders known as the Five Nations loyal to the English Crown. The West Indies were none the less harried by French privateers, who sank the Islanders' ships and interfered grievously with their food supply. It was vital that the lines of communication should be kept open. As on the Mainland so in the Islands each acted independently in terms of its own entity, and all were swayed by jealousy and local disputes.* The Mother Country was herself too burdened with the wars of William to send much assistance. In their divided state it is strange there was no great catastrophe.

In 1693 an attempt was made to strike a serious blow at France. A fleet under Sir Francis Wheler reached Barbadoes, but with that fatal procrastination that even in those days marked official transactions, it sailed too late, and through disease and gross mismanagement proved an entire failure. The ignominious end of the expedition threw Barbadoes, Jamaica, and the Leeward islands into a panic, and severely tested the fidelity of the Indians of New York. Advices from home sensibly enough recommended the isolated units to band together in self-defence. The only result was yet further quarrels.

The means of land defence were feeble. The Islands had, or were supposed to have a White Militia, composed of servants imported from the Old Country and bound to the planters for a term of years. Various causes, William's wars, low wages and bad seasons, rendered it impossible to keep up the defences; on the other hand the Islands could not be left totally unguarded. Governor Codrington (who signed the roll for Antigua) appealed for aid. Four ships and four hundred men were sent. Barbadoes, after much shuffling among the politicians, likewise received a little help. The inhabitants of Antigua expressed their gratitude in the roll: " Having to our great comfort received from Your Majtes gratious clemency and regard a peculiar protection from our inraged enemyes, etc." and in a fuller manner, the Burgesses of James City in Virginia speak of the

*Nearly a century later Alexander Hamilton, the founder of the American constitution, experienced the greatest difficulty in organising a constitutional convention. As Dr. Murray Butler says : " In those days you could not easily persuade the several colonies to come together in conference for any purpose, lest they might, in some way, as a result of conferring, sacrifice a measure of their independence and their sturdy separateness." [Address delivered at the Hamilton Club, Brooklyn, 11th Jan., 1913.]

" signall tokens and marks of yor maties grace and favour lately received in a fresh supply of ammunition and stores of war for great guns." From Jamaica, always in fear of invasion, there is no roll extant so far as the Public Record Office goes. Barbadoes certainly had no reason to be ungrateful, since the force under Francis Russell, which had commenced to clear the seas of privateers, was called off from that useful work to bring them assistance.

On the mainland matters were little if any better. In 1689 the English in North America exceeded the French in number, so much so, that Louis endeavoured to persuade William that the Oversea Plantations should remain neutral. The English King was too sagacious and refused. In reality both sides hoped much from war; the English to take Quebec, the French to conquer the Iroquois and thus obtain an open trade route from the Erie to the Mississippi. They further hoped to drive the English out of Hudson's Bay and secure the fur trade for France. Both sides fanned the ferocity of the Indians and their warriors raided each other's settlements with sickening cruelty.

At last some sense of responsibility and duty came to the Authority at home. The management of the American Colonies was then removed out of the hands of the Privy Council and entrusted to a Commission for the Administration of Trade and Plantations, viz., the Board of Trade, the Council, of course, retaining the supreme control.

Possibly hope of the new regime had some effect, but the ever present fear of France was the motive power that led to the signing of these Rolls in one of which, that of Bermuda, are enrolled tribe by tribe, the signatures of all its inhabitants.

The form of the rolls follows closely the precedents set up in the home country, but each has a character of its own. New York, by far the largest, has four distinct rolls, one for the Governor and

Council, another for the Civic Authorities, the Mayor and Corporation, a third for the Colonel, Paymasters and Officers of the Garrison and that of the inhabitants. This forms a directory of the adult males of the city and bears witness to its cosmopolitan character, certain Jews writing their names in Hebrew. A facsimile of the eleventh membrane of the New York Roll will be found facing page 44.

The Roll of the Dominion of Virginia adds a long address to the more formal oath and loyal declaration. In their address it expresses, as already noted, the gratitude of the colony for munitions of war, and also for the acceptance of funds raised for the aid of the defence of New York. The Virginia Rolls have been photographed and form the frontispiece to the present volume.

Antigua has a roll for the Governor and Assembly, another for the Military Authorities, and a third which enshrines an Address of Gratitude from the inhabitants.

Nevis has a roll which embraces all sections of the community and an address.

Montserat has two distinct enrolments, the Governor and Council, and another for the Militia.

St. Christopher's single roll includes the principal inhabitants.

The Bermuda roll is a copy only, the original being retained in the Secretary's office in the island. It includes the signatures of the Governor, Council, Assembly and inhabitants, the latter signing under their tribes.

Inasmuch as many emigrants from England reached the American plantations by way of Holland it is quite appropriate to append the rolls for the City of Dort, Rotterdam, and the Hague. The short Rolls for the English residents in Malaga and Geneva are also added.

Careful study will reveal the value of the documents, not only as a contribution to history, but as an important aid to genealogical research. It will be noted that persons bearing names well-known throughout the civilized world were even in these early days occupying important positions in New York and Virginia, and doubtless standing firm for liberty to their everlasting honour.

As Doctor Murray Butler well said, speaking to an audience at Albany, New York, 15th June, 1915, but with an application equally extending over the whole of the " Plantations " :—

The people of the State of New York inherited and brought across the sea the political and social institutions of the seventeenth and eighteenth century England. The Constitution of England was their constitution, and into the rights and benefits of Magna Charta they entered as the lineal descendants of those free men of England to whom those rights and benefits had been assured forever. When New York was still a colony, Chatham, replying in the House of Lords to the Marquis of Rockingham's speech on the State of the Nation (January 22nd, 1770), said: ' The Constitution has its political Bible, by which, if it be fairly consulted, every political question may and ought to be determined. Magna Charta, the Petition of Right, and the Bill of Rights form that code which I call the Bible of the English Constitution.' These three great documents mark the progress of the struggle between the barons and the people of England with the Plantagenet, the Tudor, and the Stuart kings, through which struggles the government of England was gradually transformed from a feudal monarchy into a democracy in fact, with an elective kingship and an aristocratic social system.

In the Association Oath Rolls of 1696 we have the rare spectacle of a nation sadly divided by jealousies of various kinds, but fired to a tremendous enthusiasm for the preservation of their solid hard-won liberties, and in the Rolls herein transcribed we see the reflection of this spirit in the young colonies, despite the well-known fact that they were founded rather as secessions than as expansions of the Empire.

I have thought well to transcribe these Rolls and to indicate a sketch of some political conditions in the English Empire of that period for the help of my fellow students in all centres of Anglo-American historical research.

THE ROLLS.

[*P.B. Assoc. O.R. 465. Endorsed* :—]

ASSOCIATION SIGNED BY THE GOUVERNOUR COUNCILL AND ASSEMBLY OF

BARBADOS.

Barbados.

Whereas . . . . Crown.

May the 14th, 1696.

T. Russell.

THE ASSEMBLY.	THE COUNCILL.
Tho. Maxwell	John Hallett
James Colleton	Fran Bond
Abell Alleyne	John Gibbes
Thomas Merrick	Edw Cranfield
Jon Lesled	John Farmer
John Brome	Richard Slatter
Wm. Fortescue	Geo : Lillington
Wm. Allamby	Geo. Andrews
Jona Downes	Jno. Bromley
John Cousings	Wm Sharpe
Robt. Bishop	Pat : Meine
Wm. Holder	Job Frere
William Dotin	G. Payne
Will. Wheeler	Clerk of the Council Overte
John Bishop	
George Peers	
John Wiltsheir	
Tho. Maycock	
John Waterman	
John Waterman jun.	on dorse
Wm. Cleland	
Wm. Rawlin offcr the Assembly	Ro : Hooper attorn gen.

[*P.B. Assoc. O.R.* 466. *Endorsed* :—]

ASSOCIATION OF THE CLERGY OF

BARBADOS.

R Sept. 18 1696

No. 466.

Barbados.

Whereas . . . .

. . . . the Crowne

May the 14th 1696.

Ben : Cryer

Rand : Vandreye

William Ball

Ben : Callow

Henry Deane

C. Irvine

Samuel Ebriane

Gilbert Ramsay

Benja Hargrave

Jno. Milward.

[*P.B. Assoc. O.R.* 467. *Endorsed* :—]

ASSOSIATION SIGNED BY THE OFFICERS OF HIS MTIES GOVRS REGIMTS AT

BARBADOS 39.

Barbados R. Sept. 15 1696.

No. 467.

Whereas there has been . . . .

. . . . the Crowne

Rand Vandrey Rector

John Mossley	T. Russell
R. Hopson	Tho : Garth
John Wilde	Jonathan Langley
G. Payne	Jos : Crisp
John Sharpe	Robt Bishop
Joseph Woodrooffe	
William Boyle	

[*P.B. Assoc. O.R.* 468.]

VIRGINIA.

October 20th 1696

FORASMUCH as it is notoriously manifest that there hath lately been an horrid and detestable conspiracy of Papists and other barbarous and bloody traitors in ye Kingdome of England, to take away his Maties life by assasinateing his Royall Person, to the end of an intended Invasion from France for the subversion of the Religion Laws and Liberties of that Kingdom and in that of this and all other his Maties Dominions might be thereby the better facilitated

We whose names are hereunto underwritten the Burgesses assembled at James City in his Maties Dominion of Virginia do heartily sincerely and solemnly profess testifie and declare that his present Maty King William is our Rightfull and lawfull King and we do hereby mutually promise and engage to stand by and assist each other to the utmost of our Power in the supporting defending and keeping this Governmt for his Maty against the late King James and his adherents and if it should so happen that his Maty should come to a violent or untimely end (which God forbid) we do hereby protest and declare that we wil be enemies to all persons that have been his Enemies and also that we will unite associate and assist each other in the defending and keeping this Dominion for such Successor of his Maty as the Crowne of England shal belong to according to an Act made in the first year of the Reign of King William and Queen Mary. Entituled an Act declareing the Rights and Liberties of the Subject and setling the Succession of the Crown.

Robert Carter	John Thorowgood
William Byrd	Wm. Leigh
Wm. Bassett	Joshua Story
Gideon Macon	Jas. Benn
Ben. Burrough	John Giles

Mord. Cooke
Richard Haynie
Rodham Kennet
Benja. Harrison
Jon. Thompson
Alexr. Spence
Isaac Allerton
Hen : Luke
Michael Sherman
George Heale

John Brasseur
Tho : Hodges
Thomas Mason
Anthony Armistead
Willis Wilson
Dudley Digges
Rich. Whitaker

Jno. Washbourne
Rich : Bally
Alexr. Newman
Samll. Travers
Char. Goodrick
John Taylor

Matt. Kemp
Robert Dudley
G. Mason
John Withers
Jno. Battaile
Tho : Ballard
Wm Waters
Wm Sherwood
Wm Randolph
Hen : Jenkins
Jno Custis
Tho : Jordan

TO THE KING'S MOST EXCELLENT MAJESTY.

We yor Ma'^{ts} most Loyall and dutifull subjects the Burgesses now assembled at James City in yor Maties most ancient Colony and Dominion of Virginia, having taken into our serious Consideracon how much yo'^r Ma'^{ts} Subjects here stand obliged for yor Royal care from time to time extended to this yor Dominion and in a more particular manner for those signall Tokens and Marks of yor Maties Grace and favour lately received, in a fresh supply of ammunition and Stores of War for great Guns, in yor Maties gracious acceptance of the money given by the Assembly for the Assistance of New york and thereupon dispenceing with the Quota of men commanded for that service, in the informacon wee have had of the timely and most happy discovery of the late horrid conspiracy against yor Sacred Person and Intended Invasion of yor Ma'^{ts} most Ancient Kingdome of England, in the notice we have been given of the preparations of the French against yoᵣ Maᵗⁱᵉˢ Plantations this way, to the end wee should not be surprised and lastly in yoᵣ Maᵗˢ most gracious assureance of assistance according to our Exigences in all humble and thankfull manner acknowledge yoᵣ Princely Bounty Goodness and Care for and towards the peace defence and security of this yoᵣ Maᵗⁱᵉˢ Dominion, and as we do hertily congratulate yoᵣ Maᵗˢ preservation and deliverance from the barbarous designes of yoᵣ Enemies, with an utter abhorrence and detestacon of such designes so we do daily offer up our Prayers to Almighty God for Continueance of his mercies in blessing us with a long and Prosperous Reign of yoᵣ Maᵗʸ over us, and on our parts do unfeignedly promise upon all Occasions to expose and lay downe our Lives and fortunes for the support and defence of yoᵣ Maᵗʸˢ Interest and Laws and in the Opposition of yoᵣ Enemies, and as a Testimony of our Loyall and dutifull inclinations and sincere affection to yoᵣ Maᵗʸ (according to the Practice of yoᵣ Maᵗⁱᵉˢ Loyall and good Subjects at home) we have every of us entred into an association, which we have prayed yoᵣ Maᵗⁱᵉˢ governor here to assist us in getting presented to yoᵣ Maᵗʸ and in all humility beseech yoᵣ Maᵗⁱᵉˢ gracious acceptance thereof.

October the 20th 1696

By order and in behalf of the House of Burgesses

ROBERT CARTER
Speaker.

[*P.B. Assoc. O.R.* 469. *Endorsed* :—]

GOVERNOR AND COUNCILL OF

NEW YORK PROVINCE.

Whereas . . . .

. . . . Crowne. Dated at His Majestyes Fort in Newyorke in America the 26th day of May 1696 Anno Reg. Reg. Willi. Tertii Anglie &c viij°.

Ben. Fletcher	W. Nicoll
Fredrych Hyple	Wm. Penhorne
John Lawrence	Chid. Brooke
S. Cortlandt	M. Clarkson Secry
G. Miniveele	David Jamison
N. Bayard	Comitie
Caleb Heathcote	

TO THE KING'S EXCELL^T MAIESTY.

The humble Address of y^r Maiesties Capt. Generall and Govern^r in cheife and the Councill of you Majes^{ties} Province of New Yorke in America.

WEE your Majesties most humble most Loyall and most Obedient Subjects deeply sensible of the great and Good providence of Almighty God   Lately manifested to all y^r Majesties good People in y^r most gracious and wonderfull deliverance of your Sacred Person from the horrid and detestable Conspiracy of y^r Enemys do heartily congratulate the same and do dayly offer up our Prayers to Almighty God for the Preservation of y^r Majesties Person so frequently exposed to Danger for the Preservation of our Religion and Liberty and to Grant unto Your Majesty long Life a Victorious and Happy Reigne.

Newyorke Janry^e 11th 1696.

	Ben Fletcher	
N. Bayard	Frederych Hypple	S. P. Cortlandt
W. Nicoll	Caleb Heathcote	G. Mininveele
Chid. Brooke	John Lawrence	Wm Penhorne

[*P.B. Assoc. O.R.* 470.]

MAYOR RECORDER ALDERMAN AND COMMONALITY OF THE
CITTY OF NEW YORKE.

WHEREAS there has been a horrid and detestable conspiracy form'd and carried on by Papists and other wicked and Traiterous Persons for assassinateing his Majest[s] Royall Person in order to encourage an Invasion from France to subvert our Religion Laws and Liberty Wee the Mayor Recorder Aldermen and Commonality of the Citty of New Yorke doe heartily sincerely and solemnly Profess Testifie and declare that his Present Majesty King William is Rightfull and Lawfull King of these Realms and wee doe Mutually Promise and Engage to stand by and assist each other to the utmost of our Power In the Supporte and Defence of his Majesties Most Sacred Person and Government, against the late King James and all his Adherents and In Case his Majesty come to any violent or untimely Death (which God forbid) Wee doe hereby freely and unanimously Oblige our Selves to Unite, Associate and Stand by each other in Revenging the same upon his Enemies and their Adherents, and In Supporting and Defending the Succession of the Crown According to an Act made in the first year of the Reign of King William and Queen Mary Intituled an Act Declaring the Rights and Liberties of the Subject and settling the Succession of the Crown.

Will. Merrett Mayor
Ja. Graham Recorder

Rip Van Dam
J. S. D. Spiegel
Johannes Hardenbroeck } Assistants
Martin Clock
Jan Eirwets
Rich Ashfield High Const

John Tuder Vic. Com.
Will Beechman Alderman
J. S. Cortlandt
Brandt Schuyler
Robert Darkins
Jacob Boeleuz
Gerard Douie } Aldermen
Will. Sharpas Cl.
Ebenezer Willson Chambr
Edw Buckmaster

34

PEIJMRS & CONNENOLL.

Charles Lodwik. L'. Col.	Ja. Emett
R. Bruijir Capt.	Will. Morris
Cornelius De Peyster	Robert Walters
Andrew Teller Junr	John Merrett
Theunis de Keije	Law : Reade
Tho. Monsey	Jeremiah Tothill Capt
Stephen de Lamdy	John Barbirie
Jacobus Dekeij	N. W. Stuijvesant
Tho. Wenham	Gea : Rescarrick
Benj Phipps	tomas torriner
Dirck Vanderburgh	John Clapp
Isaac de Foreest	Strent Schuyler
Ja. D. Peijster	Dan. Honan
J : V : Riemer	Cha. Young
John Tuder Junr	David Jamison
J. Jansen	
Sam : Baijard	
Johan Cortlandt	
Tho : Clarke	
Johannes Kist	
Rob. White	
Robt Lurting	

Thomas Smith X

Peter Janyn Codchhout X

David da Jover X

Peter Van Tilborge X

Henry Crosley X

Dirck Slejch X

Willm. Peetshen X

Nicholas Dumaresq X

R. White

Gustavus Adolphus Horne

Johans Pietersen X

Garet Bras X

Euerardus Bogardus

Handrijck Meslaer

John White

Salama Fedrijxdsboog X

Lenard de Geor X

Gaybeit Van bergh

James Burtell minister

Alexander Lamb

ijoh Hoonuick

jooseph Waldron

Daniel Waldron

Harmon degraell

Hendrik Van Oosteom

Adam Carlis

Jacobus Kroeger

Daniel Butts

Samuel Waldrom

Jacob tounear X

Harmon Arson X

Hendrieck Vanobliens

John Smeth X

Wilter Oblienis

Isaci Oblinins

Johannes Meijer

Cornelis dijckman

Jan Dijckman

Lourans Johnson X

Albert Son

Abrm de lamotaine

Fiminis Salsen

Isac de ancette ?

Andriis Klause ber-windt

John Boult

Jerenimus Barhight X

Johanes Cornelosen

Adolf Mirde X

P. Chaigneau

Abriam Abrainzi

Jacob Moens

Jan Delamontanye

John Smart

Joseph Bueno

Isaque R. Maxquez

Arcut Fredrickson

Roger Basier

Willem Teller

Jacobus Vangiesert

Garret bras

David Provoost

John Stone

Joseph Horton

George Reeuelle

John Geddes

Jacob van Noorstrant

William Green

Richard Moore

Abraham bochee

Iberdt Heereman

Arian Van Schick

Johanas Tomason

Nicklas Gisbord   X

Jacob Couenhouen   X

Davjd Van de Weil

barnet van tellbera   X

Tenes Edeson   X

abraham Johnson   X

Marin Roleson   X

John Duveer   X

Tiidrige Cornelyse   X

Gerrit Viele

Samuel Taylor

Pietur vandenrjen

Thomas Sanders

Heny Crab   X

Jan Flerisburgh

Clemment Elsewert

Gerret von Hersse

Jacobus Colijer

Wulfam Shuttlewood

Jan Fause

Wyllem Elswert

John Petsy

Isack Vreden burgh

ijohanniis van Strab

Corne Eclelson   X

Cornels Poulesdon   X

Jacob Salamons

וולף בר ישק

[Wolff son of Saqui]

Jan. Mead.

Jan Cierssen

Casper Mebije

William Thomes X

Jaac de Harrietelt

Jan Nagel

Johannes Waldron

Bastean Mackelson X

John Bogarte X

Pieter de Riemer

Johannis Poiileise

Johanis Van de Water X

Gerrit Onckelbogh

Jan Sighons

Philip Minhorn

Tho Lewes

Margell Johnson X

Thomas Gleaue

Cornelius Lodge

Abraham van Laer X

Abram van de Len X

John Coolley

Jacob Phenix

John Canelier Senior

Edward Cox

Joost Leijn-sen

Andrew thomson

Daniel Lambert

Mich. Hawdon

Albert Johnson X

John Fredreck X

James Algeo

Josuf Daue X

Zebulon Carter

John hutton

Joost Adelsen

Theophellus Elsworth

Pieter Jansen Langedijck

Philip Darveg X

Jacob Hason X

William White

Boullen Clason

Jacob Provost

George Rayeson X

John van Horne

John Rayke

Francis Anthony

Joris Elsweit

Adoelvijs haer den Broeck

Derck boogerdt

Goosen Coster

John browne X

John Broadg X

James Hunt X

Walter Hooper

Tho Lance

Johannis Joostn

Coornelis ijoosten

Jacob Fredrycksen

Gerret Heijer

Jean Dubois

John Dublet X

Jno Daniell

Henry Dally

Albert Leendert

Abraham Nijttersale

Meijer Martens

Abraham Rozeau

A : Appel

René Rezeau

Ben. Hering

Joost Paldwick

Benj : Aske

John Baker X

Richard Pangborn

Tice Suruec Survac Tice

Jacob Swaen

Laurin Cortfin

Antony Cosart

Johannis Trocroft

Worter heyer

Nicolas Delaplaine

Jacobus Salomons

Agosten Vanderck X

Danell Coford X

Isack Brasker

Johannis Hardenbergh

Petter Mellot

Johannes Hooglandt

Evert beijvancke

William Dobbs X

William Rooseboom

Will : bogard X

Phillip Coninghame

Jacobus goelet

Winsendt delamontanyee

ijsach Vandursee

Henry Baignoux

Will : Williams

Francis van Consissven

Jasper Vissnet

John Kissane

Jan Williemse

Pieter Williemse

P. L. Grand

William White

James Symes

Robert Hawkins

Daniel Mesnard

C. Viele

James Colly

John Williams X

François Hullin

Augustus Grasset

Wm Jackson

Isaac Lenoir

Jan Van Shade

Jno Basford

Peter King

Peter Morin

Johannes tidbout

Hendrijch Meijer

Jean Bouger

John Contesse

Adreijan Hooghlant

Frans Gde brandt

J. Cartier

S. Vincens

Pieter Mijer

Johannis Teller

Poulus Turck junier

Jesse Kip

Justus bosch

Joggem Rocliffe X

Simon Romegn

Pieter Ryckman

h. Jourdain

John Pieroo

John peterson Melot

Bartholomius Vonek

Marfan bechman

Jos Blydenburgh

Wm Cothonneau

Peter : D : Mill

Hendryck Bosch

Samuel Bosch

Jean David

Thomas Vilinne

Andries Marschalek

Ouziel van Swieten

Gerrard Banker

Nath. Marston
Jaques Houtijn
Nicolaes de Foreest
John Cruger
Thomas Roberts
David Provoost Junr
Nicholas Feilding
Fra. Chappell
Thomas Swine X
Johannes Schenk
Tho Robeard X
John Beeckman
Pieter Bogardt X
Anthony Farmer X
William Haines
Abraham Metseler
Jacobus Kierstede
Gabl Ludlowe
René Rezeau
Joseph Yard
Arthur Frithy
Joseph Yard
John Miller X
Abraham Splijnter
Isaac Seloover
Jacob bennet
William Barton
Laurens Van Berck

Humph Tregenna
Francis Corles X
Richard Willett
Andrai Cardell
Adam Wallis
Jan Van Varick
Henry Kemble
Thomas Ives
John Hede X
Jno Crooke
Coenradt teneijck
Saemel beeckmaen
Thomas Milton
Pieter ijadcobs
Abraham van Gelder
Matthew Ling
John Yeates X
Samuel Staats
Joh : Groenendijch
משה הללוי
[Moses Levy]
Bestrand Shuvgivean
Pieter Aday
Ezechiel Grazillier
M. Clarkeson
David Valentijn
James Evetts
Coeurades Vanderbeeck
Paulus Vanderbeeck
Barent Janse bos.

Abraham Jouneau
Andrew Lauran
Philippe Jouneau
Jan du Mortier
James Spencer
G. Berthonneau
Jerret Roos
Jacques Morices
Andri Anriand
Jean David
John Watson X
William Churcher
Franceys Cooley
Albartes Ringo
Hartman Wessels
Pieter Sqankam
Abraham Brashet
Joannes Kertbye
William Whitty
Jacobus Von Rollegom
J. Vincenz
Heijman Conijuck
James Isells
Pieter Bontekoe
Daniel Joüet
Gerrit Hollaen
ijonthan proust
John Unsell X
Elie Papin
tobijas Stontenburgh
B. Baijard

D. Poutreau
Coele Voelen
Johannes Wessels
Sam Burtt
Evert Van Hoeck
Shoerts offerts offert Shoerts
Abraham Vander Hüel
Abraham Van Gelder
Isaack Kiss
Petrus Kop
Buerge Meijnderlen
Edward Hunt
Hendrych tendyck
Jean le Reux
Willem Appel
William Hellakers
Barent Rijnder
Richard Greene
Peiter Chaber
Arthur Strangey X
Charles Crommelin
James Spencer
Daniel Crommelin
Jehan Marckeeer
Isaac de Mill
Pietre Newkerke
Isaac Stevens
Gerret burger
Richard Calgo

Jacob Cool

Jarvis Marshall

Samevel Mijnerdt

Jno Theobalds

Obedj Silby

Cornelis Vander Birch

Pieter Janse bos

Frans Wesels

Isaac Anderson

Thomas Hooke

Fijmen van Borchsun

Richard Potter

Frans Van der hoeck

Sharl de Niser   X

Willm Moos   X

Albertus Vanderwater

Abraham Mesijer

Pieter Janser mesier   X

Daniel Targo   X

Peter Rezeau

Cornelis Kloopper

lodowyck Vandenbergh

Tennis Tybont

Gerrit Vejnans

Isaach Abrams

Cornelis Vander Wers

Cornelis Beech   X

John Pieters bos   X

Isaac dela Montainye   X

Jacobus Van Denck

Phillip Palmer

Saul Brown

Allexsandre Morrisse

Pitre Fillieux

Jean Andriuer

Edward Graham

Simon Bourn

Andrew Faucaut

Isaac Garnier

Robert Lij

David de Roblet

Leonard Lewis

Lorrens Thomass

John Pieters band   X

Johannes Hybon

John Wendover

Harmanus Vangelder

Abraham Ketellaz

Willem Hyer

Johannijs Nijs

Hendryck Kermer

Wm Huddleston

Isaack Stontenburgh

Jacob Vandeuer X

Jacob Vantilburgh

Johannes Vangeldern

Pieter Adoth

Tho. Adams

Jacobus Vanderspiegel

James Hand X

Piet Isaakson Van Hoeck

Laoers Hendrick X

John Jeanissen X

Jurijan Klanck X

Henry Janssen Van Breevoot X

William Waldron

Jhon Hendrix Van Brevoort X

Hendrich Horder

Peter Jansen Band X

Johannes ijansen Bant

Claes Burger

Johannes burger

T. Mashol

Jan bos Langestrad

Barnardus Shardenbroek

Johanes Shardenbroek

Adraien Man

James Prefost X

Cornelis Pluvier

Thomas Wood

Abraham Van horne

Andrus Hendrecke

Dirck Bensem

Timothy Archamband

Jacob Marius Groen

David Hendricks X

Jacobus Cock X

Simeon Soumain

Jacob Mayle

J. Mol

Cornelis bulfinch

Wm Coales

John Robinson

A SHEET OF THE NEW YORK CITY ROLL, PAGE 44.
Reproduced by kind permission of the Master of the Rolls.

Geylbert van in borgh

John Hope

Gouste Bonnin

Sijmon Breese

Giles Gaudineau

Laurens Meares   X

Joannis Outwan

Johannis Venderheid

Nicolaes Roosevelt

Johannes Van Vorst

Barth le Roux

Ebert Van de Waeter

Stephen Richards

Jacobus Kiss

Tomas Luiseerts

Johannes Van haut

Jean le tourrette

Zacarie Anguei

Louis Gitons

S. Apins

Moyon

Rober Sinclair

Edward Courte

Giles Kingsfeild

Phill. Mullins

John Spratt

Isaack De Peyster

Peter Cullum

Voornaer Wessels

Jean Berthonnau

George Ingoldesby

Jan Evertsen

Daniel Pheing

Claude Brueyers

Jasper Uissefert

Stephen Jamain

Samuele Bourdet

Albert Clocke

Jean Polletreau

John Membrew   X

Ebert Duijking junior

Thomas Mostyn

Henry Coleman

Johannis Renselaer

Jan de Forsett

Jean Coesaunt

Leendert Huygen de Kleyn

Drick ten eijch

Jean le Chevalier

Everdt duijckinck

Isaac Girard

Jan Heinberdinck

Jan Vinsent

Giles Stollard

Jacob Teller

Jan Peeck

Abraham Vandewater

François le Conte

Gabriell Jourasen

Pieter Jacobs Marrits

P. D' Lanoy

Barne Hetzon   X

Cornelis de Peigler

Edward Blagger

Andrijs Mijer

Joahnnes d'honneur

Abraham Santvoort

Daniel Mer Cerrau

N. Maltsaeler

P. Slats

Paul Droilhet

Gabriel Le Boyteulz

Louis Carré

Elie Kanbert

Garrit Van Hoorn

Bon Grande

Paul Richard

John Kemble

Nycolaes Blanck

Johannes Provoost

Laur Waldron

John Hutchins

Samll Paul Dufoue

Pijeter Burger

John Van Giesen

Geleert Bertsbergh

John Ellison

Christ. Eijaenlorsen.

Elie Pelletreau

Isaac Gouverneur

Lodewijck Ders

Carston Luerse

Jan breste

Johannes Wessele

Jeremijas Westerhout

Audris Breste

Caspar Steinmets

Jürijan Bosch

Phil. La Noy

Richard Gravenor

Jan Dou

John Thomas

Enoch Hills

Will. Mosse   X

Nicolas Gamain

A. Teller

P. Belin

Jacobus Berry

Claes Andrees

Elli Boudinotes

Benjamin D. hurriett

Josué Dauid

Henderick Fillise : Mijer

Guij Kuick

Carsten Luersen

Johannes Vredenburgh

Gerret de Gracier

Hindereck Vanskiek X

Philip Bristoll

Walfort Webbe

Jan. Eckesen

Rein Wackelen bos X

ijehannes Pietersen

ijan Eckesen X

Leijies ijansse brewoort X

Philip Carder

Wonter Areson X

Bastian Ells X

Solomon Pieters X

Peter Cauelidz

Henderey bup

thomas Eckis

Woolford Weber ʌ

Jochhe Hereman

Jan Arijansen

Paulus Turke X

Gohom Jonse Von Arnum X

Walter Dobbs X

Jacob Cornelisse X

Danell Francesco X

Thomas Carrells

Will. Welch

Saccharijas Seckales

Laurence Jonse X

Robert Skelton

Adruent Webbers

Peter Luckas

Andrs Jorn

David bonnefoy

Hieronymous Schout

Resper Pieters

Cornelis Van de Huyt

George Owen

Moses Owen

Cniecklas Badcher

Geletreus Cadusien X

Thomas Robbinson

Daniell Devoor X

Eliez Fruste

Arian Hendricks Van Schaick X

[*P.B. Assoc. O.R.* 471.]

BERMUDA ALS SOMER ISLANDS
In America.

WHEREAS . . . . Wee the Governor Councill and Assembly and the rest of the Inhabitants of these Islands . . .

J. GODDARD, Governor.

Captn of the FERRY FORT
Samuel Stone

Justices of ASSIZE
  Comon Pleas Etc
Wm Peniston
Gilbert Nelson
Sam' Trott
Joseph Darrell

  of EXCHEQUER
Wm Peniston
Gilb Nelson
Joseph Darrell

  of ADMIRALTY
Wm Outerbridge
Tho. Brooke

Stephen Crow Sheriffe

Nicholas Trott Secty
& Attorney genl

Justices of the Peace
Wm Peniston
Gilb Nelson
Wm Outerbridge

Nicholas Trott J[ustice] of the
  Peace and quo

Cornelius Hinson
Jno Tucker
Francis Jones
Joseph Darrell

Lieut of the Castle
William Jones

Captn of South[amp]ton Fort
Samuel Brangman

Captn of the Queens Fort
Daniel Johnson

Captn of Smiths Fort
Boaz Sharpe

In the COMPANY OF TOUNE and PARRISH of St. GEORGES

Samuel Skinner
John Apowen
Francis Harris
John Askew
Nehemiah Mote
Samuel Apowen
Joseph Cowper
Stephen Wright
Joseph Ming
Joseph North
Edward North
Joseph Wood
William Newman
Thomas Wharton
John Hurt
Charles Higgs
William Cornish
James Burchall
John Cornish
Elias Pickker
William Garmaer }
Zachariah Johnson } Drumrs
John Stow
Edward Finney
James Croskey
Joseph Brangman
Mathew Mings
John Lightbow
Willm Addams
William Davis
Thomas Stokes
Phineas Wright
John Stirrup

Thomas Judkin
Saml Jones
Edward Hubburt
William Jordan
James Allen
Samuel Bell
Henry Tucker
Asser Sharp
John Askew
Paul Vaughan
John Higgs
William Evans
Boas Bett
Boas Bell
John Fox
Francis Jones
Joseph Askew
Sam' Briggs
Benj Phillips
Thomas Sturrup
Will Forster
Saml Sharpe
Zachariah Jones
Daniel Johnson jun
Robert Burchall
John Cornish
John Harloe
William addams
Benj : Fox
Thomas Wood
Samuel Whiteing
Samuel Wood

John Tucker Captaine

Daniel Tucker Leiutent

Leonard White Ensigne

James Wright 1st Sergeant

Humphrey Burchall 2d

John Bellinger 3d

George Dew

Samuel Stone

John Tucker

John Stringer

Thomas Phillips

Alexander Smith

John Welsh

John Davis

Thomas Stowe

John Harlow

William Newman

James Burchall

John Higgs

John Middleton

Edward Middleton Sen

Thomas Cooper

John Rallins

Zachariah Briggs

Willm Croskeys

Daniel Jones

John Briggs

Experience Fox

David Ming

Roger Browne

Samuel Jordan

Joseph Ming

James Rice

Charles Apowell

Cornelius Levens

John Hilton

Ephraim Wright

John Burchall

Samuel Mills

James Burchall

Jonathan Ming

George Tucker

Joseph Wright

John Jones

William Laffing

Henry Smith

Nathaniel Mills

Nathaniel Robberds.

ASSEMBLY and COUNCIL.
Representatives of Tribes.

ASSEMBLY

Cornelius Hinson ⎫
The Marke of (R.H.) ⎪ For
Richard Hauger ⎬ Pembrook
David Whitney ⎪ Tribe
John Richardson ⎭

Nathaniel Sruddun ⎫ For
James Darrell ⎪ Pagitts
Sam Daffye ⎬ Tribe
Daniel Keele ⎭

Hen Tucker ⎫ For
Benj Wainwright ⎬ Warwick
William Basden ⎭ Tribe

John Dickenson ⎬ For Southampton Tribe

Jonathan Tucker ⎫ For
Daniel Burges ⎪ Sandys
Nathaniel Meratt ⎬ Tribe
Thomas Burch ⎭

John Kidgell Cleric
Convent

THE COUNCIL

Wm Peniston
Samuel Trott
Wm Outerbridge
Jno Tucker
Richard Veniston
Gilbt Nelson
Charles Minors Cl. Concl

THE ASSEMBLY
John Gilbert Speaker

Daniel Johnson ⎫ For the
Daniel Tucker ⎪ Town
Samuel Stone ⎬ of S' George
Nicholas Trott ⎭

John Peasley ⎫ For
Samuel Hubbard ⎬ Hamilton
Joseph Cox ⎪ Tribe
Rowland Greatbatch ⎭

Samuel Harvey ⎱ For
Moses Knapthorn ⎰ Smiths Tribe

Samuel Sherlock ⎱ For
Beniamin Stow ⎰ Devonshire Tribe

A true coppy of the Orriginal
Recorded in the Secretarys office.
CHARLES MINORS, Secty.

BERMUDAS. Enrolment by Tribe.

COMPN IN HAMILTON TRIBE

William Cox
William Penston, Coll
Anthony White Lt Coll
Mich Burrows Captn
Samuel Hubbard Lieut
William Stone Ensign
Timothy Pindar }Sergts
Malachi Hall
Thomas Harford Genl
Thomas Hall
Richard Stafford Genl
Edwin Stone
Joseph Allen
John Seare
Sam X Prudden
James Wilders
Joseph X Cox
Benjamin Low
William Smith
Edmd Mallory
John Wells
Isaac Davis
Daniel Jones
John Hawkes
Daniel Hubbard
John Hamond
John Trott
Samll Trott
Walter Turner
John Outerbridge

Joshua More
Jeremiah Burges
Joseph Packwood
Jeremiah Fiklin
John Somersall
Stephen Newbold
Jadwin Downscombe
Stephen Newbold Jun
Nathaniel Yate
William Griffin
Edward North
Daniel Greatbatch
John Cox
William Outerbridge
Edward Dunscombe
Nathaniel Dunscombe
Freeman Stone
John Peasly
John Tucker
Thomas Outerbridge
John Willis
Rowland Greatbatch
Daniel Davis
Isaac Davis Sen
Simon Hall
Michael Burrowes
William Outerbridge

Recorded in the Secrys Office

SMITH'S TRIBE

Richard Peniston

William Davis

George Smith

Thomas Bostock

Charles Walker

John Dickinson

Thomas Smith

John Gilbert

Peter Gilbert Sen

John Wingood

Samuel Newton

Samuel Harvey

Thomas Smith

Malachi Gilbert

Samuel Peniston

John Righton

Edward Graysborough

Thomas Graysbury

John Peniston

Seth Place

William Gibbins

George Argent

Elias Howell

Thomas Powell

Daniel Smith

Benjamin Harvey

William Mallagar

John Somersalls Sr

Jeremiah Smith

William Davis

William Wells

Perient Trott

Moses Knapthon

Josias Smith

Anthony Smith

Samson Potter

Richard (R) Johnson

Jeremiah Newton

Thomas Collins

Thomas Righton

Daniel Wells

Seath Place

John Burt

Moses Knapton

Rowland Tongue

Michael (M) Smith

Samuel Wingood

Daniel Strogham

John Righton

George Poell

Samuel Wells

Richard Peniston

Richard Jenyns

Seth Smith

John Wingood

Thomas Attwood

Samuel Harvey Jun

John Davis igs

Robert Skinner

Samuel Somersett

SANDYS TRIBE.

William Seymour

Jonathan Borch

Jonathan Burch

David Johnson

William Harmon

Thomas Forster

Edmond Mallory

Banjamin Hinson

Shechariah Burrows

Nicholas Spencer

Aron Ward

Aaron Duryes

Josias (I.T) Tatem

Horatin Mathlin

Joseph Agle

Joseph Rivers

James Hermer

Joh Priestley

Benj Young

Samuel Kins

William Meritt

Mich. Evans

John Claxter

Joseph Burt

John X Ward

Edward Merrit

Tho. Young

Nathaniel Conyers

George Young

Daniel Burges

Tho X Burch

Hor Seymer

Jeremiah Burch

David Whitney

Thomas Burch

Painter Burrows

George Ball

Peter Boulton

George Watson

Nath. Meratt

John Pder

Robert Evans

James Verchild

Florans Pijos

Benj. Applebe

Jonathan Hill

Joseph Hinson

John Wells

William Place

William Young

Nehemiah Rivers

Daniel Hinson Jun

Jonathan Tucker

Tho. Harloe

Nathaniel Pindar

Senior Bowe

John Forster

Tho. Vadell

Tho Watson

Tho Burrow

COMPA IN SOUTHAMPTON TRIBE

John Dickenson Capt
Francis Dickenson Lieut
Samuel Newbould Ens
George Darrell Sergt
William Bryan Sergt
Edmund Evan
Charles Morgan
John King
John Perengif
John Jenings
George Tucker
John Vaughan
Robert Dickinson
Francis Cooper
John Gibbs
Stephen Painter
Thomas Jackson
Onesimus Jackson
Benjamin Todd
Marmaduke Dando
Nathaniel Prudden
Daniel Morgan
Daniel Conyers
Robert
John Darrell
William Cooper
Joseph Todd
Samuell Darrell
Joseph Pridden
Stephen Bullock
Benj Smith

John Beak
Joseph Lightburn
William Wainwright
John Tucker
Daniel Durham
Joseph Hutchins
William Nelme
Thomas Hunt
John Geell
Henry Morgan
Cornelius Mitchell
Samuel Dickinson
Coleman King
John Todd
Daniel Caron
Benjamin Evans
Nicholas Richards
Benjamin Chapplin
Charles Evans
Daniel Styles
John Rivers
William Leaycroft
John Evans
Daniel Nelme
Joseph Durham
Joseph Jackson
Thomas Cooper
Thomas Gibbs
Thomas Leaycroft
George Dickinson
Thomas Wetherley.

COMPA IN WARWICK TRIBE

William Tucker Capt
Benjamin Wainwriht Lt
John Harvey Ens
Christopher Prudden St
Henry Tucker
Henry Harvey
John Wainwright Sen
Phineas (P.A) Ashley
William Harvey
Joshua (I) Nash
John (J.B.) Brightman
George (R) Roberts
Oliver Barry
Jonathan Nelmes
Samuel Wainwright
Samuel Wells
Joshuah Nash Jun
Thomas Gilbert
Lazarus Frith
Saml Frith
George Tucker
Samuel Wentworth
Robert Beasy
William Tatem
Samuel Attwood
John Gibbs Senr
John Besey
Nathanll Frith
Joseph Harris
Nathaniel White
William Conyers
Daniel Lushier
Charles Besey

Daniel Nelmes
John Hall
Thomas Tale
Thomas Wells
Joseph Harriott
Ezckiel Conyers
Nath : [blank]
Daniel Frith
Phil X Dunscombe
Moses Ward
John Wainwright
Henry Catlyn
Benjamin Underwood
William Tucker
Daniel Gibbs
Lazarus Frith
John Wentworth
Thomas Besey
Michael Tayler
Joseph Chapplin
Seth Harvey
Benjamin Pasolus
Samuel Nelme
Samuel Nelme
Daniel Astwood
John Cooper
Michael Sales
George Tucker
Joseph Bennett
Robert Besey   X
James Mallagan
William Basd
Samuel Harvy

COMPA IN PAGITTS TRIBE

Francis Jones Captn
Peter Prudden Lt
Nathaniel Prudden En⁻ː
Nathaniel Waterman
John Thornton
Robert White
Patrick Downing
Christopher Smith
John Trimingham
James Darrell
John Bascome
Sam Daffye
Nathaniel Butterfield
John Burch
William Butterfield
Benjn Burch
David Burch
Daniel Keele
Nathaniel Astwood
Joseph Burch
Lewis Middleton
Stephen Hutchins

Tho (T) Parker
John (I) Wells
Samuel Goode
Nathaniel Fynes
Tho. Smith
Copeland Lea
Thomas Jones
Christopher Smith
Tho : Rose
Benjamin After
John Herig
Nehemiah Duncombe
James Gilcrist
Joseph Sherlock
Peter Casson
John (I) Arthur
Benjamin Thornton
Step Fines
Daniel Ross
Thomas Casson
Aron Grenway
John Seares

COMPA IN PEMBROOK TRIBE

Cornelius Hinson Cap.
John Richardson Lt.
Thomas Wood. Ens.
Richard Stamers }
Edward Seares } Sergt.
Christopher Pitt
Bartholomew Seltus
BS Tho Seares
Josiah Newnam
Richard Hanger [R.H.]
David Watkins
George Robinson
Thomas Gauntlett
Samuel Johnson
Adam Ewe
Benjaman Wod.
David Whitney
George Amey
Joseph Blay
William White
Samuel Dunscomb
Richard X Joel
Joseph Robinson
Charles Seares
Lewis Johnston.
Thomas Dunscombe
Richard Hinson
John Ingham.
Abielk Beake
Thomas Dunscombe

John Dutch
Isaac Adderly
Abraham Adderly
John Barnett
Joseph Harvey
Stephen Ingham
William Richardson
Jos X Hutchins
Tho Burges
Willm Edwards
Christopher Pitt
Christopher Allen.
Tho : [C] Covesley
Sam Smith
Edward Hinson
John Mayner
Humphrey Dobson
Edward [blank]
Joseph Stowe
Joseph Richardson
Natha : Bethell
Joseph Williams
Richard Norwood
John Stammers
Richard [R] Morris
John X Morris
John Saunders
John [I P.] Powell
John Johnstone.

COMPANY IN DEVON TRIBE

John Morris Captaine

Samuel Sherlock Leiutt.

fflo' Cox Ensigne

John Burrowes Sergt

Jonathan Watkins

Tho : Burton

William Wallington

Richard Wood

John Somersall

Thomas Potter

George Morris

Stephen Tyans

Richard Wells

Daniel Cumber

John Morris

William Williams

Joseph Read

John Burrows

George Stone

John Darrell

Edward Sherke

John Estlack

Ephraim Watkins

Paul Newbold

John Argent.

Thomas Estlack

Tho. Eastlake.

Benjamin Bowen

Tho [T] Watkins

Thomas Ap Owen

Tho Magott

Roger Amey

Hen [H] Hopkins

Paul Turner

John Harrott.

Joseph Harriott

Benj Plumer

Samuel Dill

William Righton

Anthony Williams

William Garraway

Aaron Turner

John Place

Thomas Vickars

Robert Hawkes

Peter Albony

Joseph Dill.

John Smith

BERMUDA ROLL,
Loyal Address Concluding.

WHEREAS itt is required of Us that Wee shall be true to King William: now King of England and of the English Nation: Us say to all his just and Lawfull Commands Wee Can willingly bee subject unto, not for wrath but ever for contienc saeck: And all Commands which are otherwise wheather from him or any other Wee Shall willingly and patiently suffer under them what man Shall be permitted to imposs upon us.

Wm. Wilkinson Senr

Nath' Low

Wm. Wilkinson

Tho Wilkinson

This 29th of the 7th moth 1696.

[*P.B. Assoc. O.R.* 472.]

ANTIGUA.

The Association of the Chief Governour of his Majestyes
LEEWARD CARRIBE ISLANDS.
The deputy Governour Councill and Assembly of ye
ISLAND OF ANTIGUA.

Whereas . . . . Crown.

Chr. Codrington
John Yeamans
Row : Williams
John Lucas Speaker
Robert Carden
Ja. Porter
Charles Goss
John Tankerd
Isaac Horsford
David Martin
Edward Walrond
Geo. Gamble
John Otto
Joseph Watkins
John Gamble
Abraham Swan
Daniel Mackinen
John Roe
Natha : Crump
E S. Burges

Tho : Duncombe
Edw. Byam.
H Holts
J Palmer
Jn Hamilton
Samu Martin

Bastiaen Baijer.
1696.
Rd. Cary
Commissrs.

The Association of his Excellency the GENERALL and the MILITIA OFFICERS of the REGIMENTS of the ISLAND OF ANTIGUA.

Whereas . . . . Crown.

Jno Hamilton
Samu Martin
Sam : Horne
Philemon Bird
Peter Lee
John Gamble
John Otto
Robt Martin
Thos Oesterman
R Sizar
Geo Gamble
Francis Rogers
James Nisbitte
Thomas Long
Edward Horne
Geo Thomas
Robt Donaldson
John Frye

Chr Codrington
Row Williams
William Wainwright
Edw Byam
John Tankerd
Robert Arden
David Martin
Isaac Horsford
Cuth : Black
Nath : Monk Jun
Step : Duer.
Benjn Wickham
Ja Porter
Wm Lavington
Nathal Crump
Abraham Swan
Will   Ar
Samll Parry
Jon' Duer

Giles Watkins
Christopher Knight
John Hamlin
Cornelius Voeghen
Richard Oliver
Jno Vibbs
Francis Christian

NEVIS.

The Association of the Honourable the Lieutenant Governour The Council and Assembly, Officers and other the Inhabitants of his Majesty's Island of Nevis in America.

Whereas . . . . Crown.

I

Wm Mead
Charles Pym
Jno Perrier provs/s Marshll generall
Wm Hoare
Alexd Crafford

II

Rob : Eleis
Tho : Wallwins
Nicho Anderton
Sam. Clarke
Henry Rawlings
Thomas Bowrey
Jam : Walker
John Williams
Wm Powrey
Wm Iles
Thomas Eayrs
Tho Gateward
Tho Buttler

III

Wm Ling
James Jackson
Timothy Hare
Arthur Plomer
Jno Lewis
James Rawleigh
Ja Browne
Wm Kiss
John Hallindge
John Wignall
Wm. Cheruse

Robt Lorey
John Dasent
Charles Rowland
Wm Greene
George Littman
John Edgerly
Ambrose Howard
Tho Neale

IV.

Ja. Bevon Speaker
Wm Buttler
W. Hamilton
James Ward
Jno Hanley
Tho. Geare
Aza Pinneye
Jno Browne
Samll Gardner
T. Weaver
Wm. Bates Clerk
          to ye assembly

V

Samll Gardner
Walter Symonds
Mich Smith
Wm Bury
Dan Smith
Jno Smargin
James Thynne
Rich. Abbott
J. Palmer
Jno Smargin Jun
          Clk Councll

63

NEVIS.

TO THE KINGS MOST EXCELLEN MAJESTY.

The humble addresse of the L^tt Governour Councell and Assembly of Your Island of Nevis.

WEE Your Majesties most dutifull Loyall and Obedient Subjects doe wth one heart rejoice and praise God. . . . . and humble Addressers.

J. Bevon speaker

Wm Buttler

W Hamilton

Sam' Gardner Jun

Jno Hanley

T. Weaver

James Ward

Aza Pinney

Tho Goare

Jno Browne.

Samll Gardner

Walter Symonds

Mich Smith

Wm Burt

Dan Smith

Rich' Abbott

James Thynne

P.B. Assoc. O.R. 472]

MONTSERRAT.

The Assosiation of His Majesties LEUT GOVERNOUR Togeather with the Gentlemen of the COUNCILL and the Gentlemen of the ASSEMBLY OF MOUNTSERATT.

Whereas . . . . Crown.

John Davis Speaker	T : Delavall
Thomas Attwood	H Holt
William Finch	Autho. Naylor
Joshua Fowler	Edw Earson
Paul Duncombe	Anthony Hodges
Samll Cane	J. Palmer
Geo Fullerton	Thomas Lee

MONTSERRAT.

The Association of his Majesties LIEUT GOVERNOUR, and the
OFFICERS of the MILICIA REGIMENT of the
ISLAND OF MONTSERRAT.

George Wyke
Henry Price
Jno Cochran
Thomas Lake

Geo Fullerton
William Finch
Nathaniel Harris segt
Nathall Harris Jun
Stanton Chauncy
Mathew Pond
J. Bramley

T. Delavall
Antho. Hodges
Thomas Lee
John Davis
Edw Buncombe
Jos Littell
Samll Cane
Alexr Hamilton

MONTSERATT.

TO THE KING'S MOST EXCELLENT MAJESTY.

The Humble Address of the Leiutt Governor Councell and Assembly of your Island of Mountseratt.

WEE Your Majestyss most Dutifull Loyal and Obedient Subjects doe with our heart rejoyce and praise God for the Wonderfull Deliveraunce and Preservation of your Sacred person from the Assaults of the late horrid Conspirators and shall alwayes pray for Your Majestyes long Life and happy Reigne Haveing to our great comfort received from Your Majtes Gratious Clemency and regard a peculiar protection from our Inraged Enemyes and hitherto by Your Majestyes Princely care are put in a condition to follow our Severall Imployments with chearfullness and to Eate Our Bread with more safty then ever we could formerly pretend to in the like times And we doe not doubt but you Maiestie will still Continue Your favour to these Your Poore but Loyall Subjects That our Enemyes may have no advantage over us but that we may be in a condition to serve Your Majesty Humbly beseeching your Majesty to believe that none within Your Majestys Kingdomes or Government cann or shall be more ready to expose all that's deare and Vallueable to us or enter into any Association that may bee for Your Majestyes Service and p'servation Then Wee Your Majestyes Loyall and humble addressors—

John Davis : Speaker	T. Delavall
William Finch	H Holt
Joshua Fowler	Antho Hodges Senr
Thomas Attwood	Edw Earson
John Daly.	Antho Hodges
Edw Buncombe.	J Palmer.
Samll Cane	Thomas Lee.
Geo Fullerton	

ANTEGO.

TO THE KING'S MOST EXCELLENT MAJESTY.

The humble Address of the Chiefe Governour Depty Governrs Councel and Assembly of your Island of Antego.

WEE Your Majestyss most Dutifull Loyal and Obedient Subjects doe with our heart rejoyce and praise God for the Wonderfull Deliveraunce and Preservation of your Sacred person from the Assaults of the late horrid Conspirators and shall always pray for Your Majestyes long Life and happy Reigne Haveing to our great comfort received from Your Majtes Gratious Clemency and regard a peculiar protection from our Inraged Enemyes and hitherto by Your Majestyes Princely care are put in a condition to follow our Severall Imployments with chearfullnesse and to Eate Our Bread with more safty then ever we could formerly pretend to in the like times And we doe not doubt but you Maiestie will still Continue Your favour to these Your Poore but Loyall Subjects That our Enemyes may have no advantage over us but that we may be in a condition to serve Your Majesty Humbly beseeching your Majesty to believe that none within Your Majestys Kingdomes or Government cann or shall be more ready to expose all that's deare and Vallueable to us or enter into any Association that may bee for Your Majestyes Service and p'servation Then Wee Your Majestyes Loyall and humble addressors—

John Lucas Speaker	Charles Goss	Chr. Codrington
John Tankerd	Geo. Gamble	John Yeamans
Isaac Horsford	Natha' Crump	Row. Williams
John Otto	David Martin	Tho Duncombe
John Roe		Fran : Carlile
Abraham Swan		Edwd Byam
Ja porter		J. Palmer
Daniel Mackinen		Jno Hamilton
		Samu. Martin

ST. CHRISTOPHERS.

TO THE KING'S MOST EXCELLENT MAJESTIE.

The Humble Addresse of the Lieut Generall Thmas Hill his Majestie's Lieu^t Governour and other the principall Inhabitants of S' Christophers.

Wee . . . . Addressors.   Tho : **Hill**

Bastian Branch;
Elliza Tayler. Wid.
James Whithill
Mary Tayler. Wd

St Georges.
P. Assaily
Geo. Hazell
William Hearts
David Lloyd
Edw. Manning
Cha. Mathew
John Cooke
Edwd Chaddock
Francis Pellett
John Gusly
Edward Moore
John Sadler
Henry Meoles
Charles Harper
John King

II.
T. Harmet
Ja Biskett
John Parker
Tho. White
John Roe
Geo Hackett
Matt van hollmall

Caleb Crisp
G. Papin
Jo : Dixon
D. Duchesnns

Edw' Gillard
Wm Steephens
Buretel
Westcott
Isaac Jolly
John Johnston

III.
John Esbridge
John McArthur
Hen Burrell
Mich : Lambert
John Garnett
Stephen Payne
Wm. Willet
James Wooddroft
Jno Bourryan
Sam Crooke
Jn. Ogson
Jed. Hutchinson
Thomas Bisse
Rowe Davis
Jno Hutchinson
Papin
Jonas Akers
John Oxton

[*P.B. Assoc. O.R.* 459.]

DORT.

The MERCHANTS and others Your Majesty's subjects Residing in the Citty of Dort.

V Villiers	George Gay
William Villiers	Isaac Ellis
Matthew Prior	Johan Irish
John Swinford	Joseph Kerby
Dudley Irish	Abraham Kick
Robert Gay	John Girard
Samuel Megapolensis	Henry Seymour
John Gay	John Gachon
Jn Armiger	Salomon Balmier

ROTTERDAM.

The Merchants and others Youre Majestey's subjects Residing In the city of Rotterdam.

Westmoreland
H Woodstock
Paul Rapins
Edward Browne
Samll Green
Nathaniel Makernes
Richd Boys
Robt Fowler
John Carter
Robt Loder
Samuel Palmer
Richard Church
Steven Tracy
Wm Fawcett
John Callman
Benjamin Lisle
Daniel Sadler
Ma. Kerr
M Stapylton
chas Morrison
J. Lenser
John Spademan
John Griffith
Francis Greenwood
Martine Browne
Nich Taverner
Tho. Augustine
Christopher Bernard
Joseph Milner
Edmund Peterson
John Derrix
Robt Venn

John Cossack
Samuel Anthony
N. Gwynne 1696
George Sword 1696
Randolphus Carleill
Richard Davijes
John Augustine
Joseph Lacy
James Washington
Robert Paige
Anthony Officiall
Sam' Barons
Robert Morsse Junior
John Rogers
Tho. Cooke
William Thomas
Nichas Reve
Benjamin Wand
Augustine Reve
Henry Nicholas
John Rutter
Daniell Walton
Bertie Cocke
Joseph Fowler
Wm Cooke
Hugh Collins
Richard Davis
Richd Bennet
Sam Smith 1696
Will Jenings
Samuel Kick

THE HAGUE.

Your Majesties subjects residing at the Hague.

Phil McDonald Bowie

Henry Yorkes	Jean vie
John Chambers	John Lillie
John Colbert	Richard ball
James Mercer	George Jefferson
	Thomas Harrison
	Abram Fletcher

W E Your Majestys most dutifull and Loyal Subjects, residing in the Province of Holland have thought it equally our duty to sign the above written Association, as if we were actually in Your Majesty's Realms and Dominions and we humbly take this occasion to Assure Your Majesty that we as sincerely acknowledge thend of Providence in the preservation of Your Majesty's Most Sacred Person from the late horrid Conspiracy, and that we do ardently offer up our Prayers to Almighty God for the Continuance of Your Majesty's life and the prosperity of Your Government as any other of your Majesty's dutifull and Loyal Subjects Wheresoever all the Hague this first day of May 1696.

[*P.B. Assoc. O.R.* 460.]

MALAGA.

The association and humble address of the Conscell Deputies, and other of yᵉ LOYALL FACTORY OF MALAGA in the KINGDOME OF SPAINE.

John Radburn
John Frere
Wm Hayward
Arth : Upton
Nicolas Holloway
Charles Haworth
Anᵒy Belton
Edward Barnes

George Ford
Anthony Swyermer Jun
Virgill Reynolds
Ant Reynolds
Jno Thruppe

I

GENEVA.

Whereas . . . . Crown.

Geneva yᵉ 30th August 1696.

Robert Wiltch

> Tho : Kirke Consull
> Giles Balters
> Richard Shukburgh
> Wm Fowler

On dorse. The desenters are

> Tho : Langherne
> John Scudemore
> Charles Henshaw
> Daniell Cross

INDEX OF SURNAMES.

INDEX OF SURNAMES

www.ingramcontent.com/pod-product-compliance
Lightning Source LLC
Chambersburg PA
CBHW031732210326
41519CB00050B/6308